TABLE OF CONTENTS

DEDICATION

I dedicate this work to all health workers in every country of the world and all that lost their lives to COVID-19 (*A minute silence for them.*).

ACKNOWLEDGMENT

I sincerely acknowledge all persons, whose work in COBOL language or contribution aided in a way or the other in the successful completion of this work. I also acknowledge you; my reader and I believe that you will find this material helpful and I wish you success in your coding endeavors. Happy reading and coding.

INTRODUCTION

Right from the beginning of time, the ability to think, create and innovate has always been one of the seeds sown into mankind by the divine. The invention of the computer orchard in a new form of creativity into man, that one with access to the computer can create something worthwhile while relaxing at home. The ability to write a computer program is one of the most cherished skills in the world and a lot of programming language is at the disposal of anyone willing to learn, how to code. However, in this material, our focus will be on the COBOL programming language. We will look at the makeup of the language, the shortcomings, etc. while ensuring that all the information is delivered in a head-first manner, for easy understanding for anyone. With that said, let's dive in.

Chapter 1

- BRIEF HISTORY OF COBOL.
- COBOL KEYNOTES.
- COBOL STANDARDS.

The acronym COBOL stands for *common business-oriented language*. It is a computer programming language that looks like an English sentence/word primarily designed for business-oriented programs. It is a procedural and imperative language and 43yrs (1959 - 2002) after its inception it became objected – oriented.

BRIEF HISTORY OF COBOL:

In the late 1950s, the rising cost of computer programming, spark a meeting at the University of Pennsylvania which was organized by Mary k. Hawes in 1959. At the meeting the DoD (Department of Defense) agreed to sponsor the development of a new programming language which will help them in data processing, being that Fortran programming language, lacks such ability.

On May 28 and 29, 1959, a meeting was held at Pentagon, there three committees where setup; Short-term committee, intermediate committee, and long -term committee, they were charged with their respective duties. On June 4, the committee was renamed to CODASYL – *committee on data system language*. A few months later, another meeting was held to discuss the name of the language, names like INFOSYL – *Information System language*, BUSY -*Business System*, COCOSYL – *Common computer system language*, COBOL– *common business-oriented language,* etc., were suggested. COBOL was later chosen, and Bob Balmer claims that he was the person that suggested it. Between September 1959 and January 1960, the Conference/Committee on Data Systems Languages (CODASYL) submitted the specification of their research, to the executive committee of the CODASYL which was approved and forwarded to the government printing office for it to print as COBOL-60. Months later, logic flaws in

COBOL-60 led to the development of COBOL–61 and subsequently COBOL-65 and so on.

In 1968, ANSI – *American standard institute*, standardized COBOL– 68 and it became the bedrock for further development of the language.

KEYNOTE ON COBOL

1. It was designed for programming business computer programs in industries like finance, human resources dept. etc.
2. It also aims at the easy development of data processing programs.
3. It aims to be easy to learn and understand.
4. It aims to support multiple system architectures.
5. It aims to be portable and fast on different platforms.
6. It evolves consistently with modern technology.

COBOL STANDARDS

COBOL at the early stage was not standardized making it easy for developers to easily modify its core structure. however, as from 1968 COBOL was standardized. And COBOL- 68 became the bedrock for further COBOL development.

Chapter 2

- APPLICATION OF COBOL.
- ADVANTAGE S OF COBOL.
- COBOL Shortcomings.

Right from the inception of the language, its purpose has always been multidimensional which includes;

I. Curbing the rising programming cost
II. Integrating the ability to program data processing programs etc.

Its multi-dimensionality made it withstand the test of time, in that it is still been widely used by much big cooperation even today.

APPLICATION OF COBOL:

the language has been optimized to be a do it all for business organizations in need of programming language that can aid them in their data handling operations. Therefore, it is an application that is mainly business-focused. the organization's like financial organization, human resource organization, government agency makes noticeable use of the language.

ADVANTAGES OF COBOL:

- It is a self – documentary.
- It can handle huge data processing tasks.
- The newer version is compatible with projects created with old versions.
- It has effective error handling techniques.
- It is a high – level language.
- It is easy to learn.
-

SHORTCOMINGS OF COBOL:

- It is not suitable for web application development.

- It is referred to as an outdated language.
- The insufficient number of COBOL programmers.
- The computer science department rarely offers its class.
- Its syntax looks a bit verbose.

Chapter 3

- ## INSTALLATION & SETUP.
- ## KNOWN COMPLIERS & IDES.
- ## PROGRAM STRUCTURE.

INSTALLATION AND SETUP

- ### COBOL ON WINDOWS:

There are many ways of installing, coding and compiling COBOL programs in windows. however, below is the easiest approach one could find.:

- Go to: www.launchpad.net/cobcIDE/+download

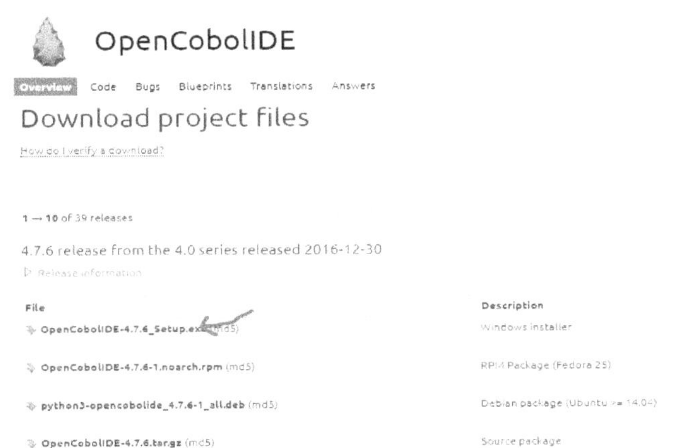

Download and install the program and you are good to go.

First Program:

- Open the installed program (OpencobolIDE) and click on the new file:

NB: On the pop-up message box, enter the program name, select a file path, and click Ok. The IDE will open with a default 'Hello World' Program code

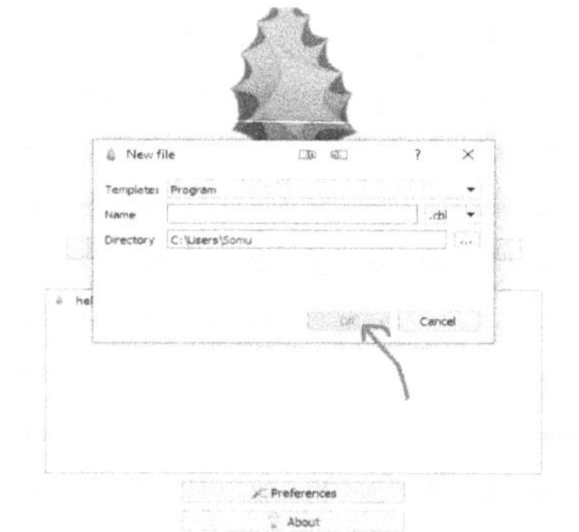

NB: To execute the code, click on the hammer icon, to first compile the code and then click on the green play button to execute the code.

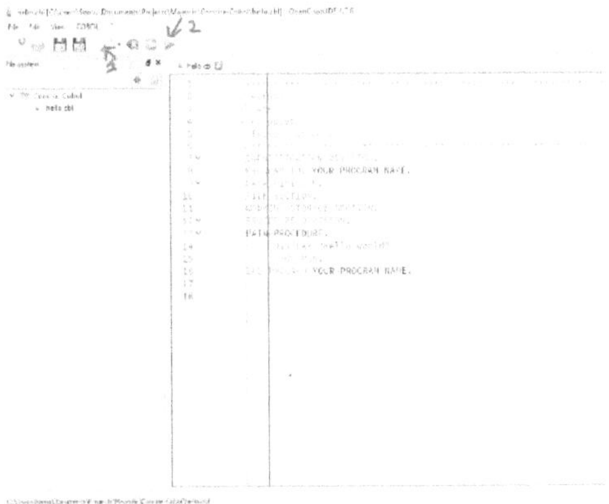

Now you can start creating programs.

- ## COBOL ON LINUX:

The easiest way to compile and run COBOL code on the Linux operating system, according to the research and experiment carried out in the process of writing this book is as stated below:

- Go to: www.launchpad.net/cobcIDE/+download and download the .deb file.

- Install the.deb file either by double-clicking on it or using the Linux command line dpkg.

Using Command Line:

Type ctrl + alt + T, to open the terminal. Then enter;

sudo dpkg -i path_to_deb_file

After the installation, open the application menu and double click on the *OpencobolIDE* icon to open the program and start coding.

KNOWN COMPILERS AND IDES

For the easy development of COBOL applications individuals, as well as organizations have developed different compilers and IDEs (Integrated development environment). For smooth COBOL programming, these compilers and IDEs are as follows;

COMPILERS:

I. **IBM COBOL compilers** – it is offered by IBM. It has an enormous amount of releases and versions. This compiler is mainly for mainframe systems.

II. **Fujitsu Netcobol** - it is offered by Fujitsu. It supports Windows, Linux, and Solaris operating systems.

III. **GNU COBOL** – it is offered by the open-source community and it is known to be the most popular of all open-source COBOL compilers. It supports Windows, Linux, and Mac Os.

IV. **Microfocus COBOL compiler** – is offed by the micro focus international plc. It supports Windows and Linux operating systems.

V. **Tiny COBOL** – it is another COBOL compile offered by the open-source community. It is based on the COBOL - 85 standards. It supports Windows, Linux, and Unix operating systems.

<u>COBOL IDES:</u>

I. **COBOL – developer studio:** it is a COBOL IDE which is an eclipse – based development environment. it comes with tons of features like:
   - COBOL code debugger.
   - Real-time code profiling.
   - It can be used on Windows and Linux operating systems.

- Code syntax highlighting.
- Code colorization.

II. **OpencobolIDE:** it is a COBOL IDE offered by the open-source community. It is lightweight and based on the GNU COBOL compiler. It is offered with features like:
    - Auto – code indentation.
    - COBOL code completion.
    - It is a crossed platform i.e. it can be operated on Linux, Windows, and Mac operating system (OS) systems.
    - COBOL code syntax highlighting.
    - It has a navigable tree view of the COBOL section.

III. **Is-COBOL IDE**: it is another eclipse – based development environment for COBOL. It comes bundled with features like:
    - Syntax highlighting.
    - Integrated debugger with remote debugging ability.
    - Real-time code syntax checking.
    - It is cross – platformed.

- It provides a consistent development environment.

Apart from using IDEs like the above-listed COBOL program can be written with the help of extensions on a text editor like;

I.   Visual studio code
II.   Sublime text
III. Slick-edit COBOL editor

PROGRAM STRUCTURE:

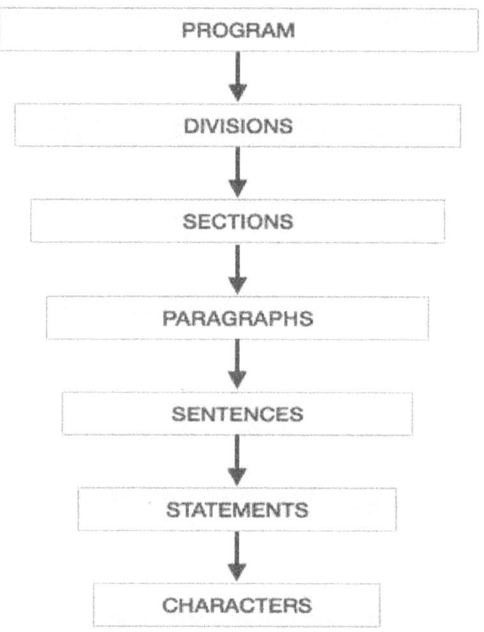

As the figure above indicated, in COBOL programming, there is an up-down, hierarchical, syntactical, and logical structure of programming. These structures include:

- Divisions
- Section
- Paragraph
- Sentences
- Statements

- Characters

It is not mandatory that all the structures must be present for the relationship to exist, for it is already pre-code into the COBOL core architecture.

UNDERSTANDING EACH STRUCTURE:

▪ <u>DIVISIONS</u>

it is a code block, which contains one or more sections or set of sections, sentence, or block of statements. It begins with the name of the division and ends at the beginning of the next division or the end of the program. Divisions are coded within the ***AREA A***.

<u>COBOL DIVISIONS</u>

I.   IDENTIFICATION DIVISION
II.   ENVIRONMENT DIVISION
III. DATA DIVISION
IV.   PROCEDURES DIVISION

I. IDENTIFICATION DIVISION: it is an essential COBOL division. It provides the program's name, that information that uniquely identifies a COBOL program. Entries located in the identification division includes;

- Author ------ the programmer
- Installation ------- the location of the installation
- Date – written ---- the date when the program was written
- Date – compiled – the date of the program first compilation
- Security dependences ---- it indicates the program's level of confidentiality

The keyword "program-id" is a mandatory entry that comes immediately after the identification division. It specifies the program name and can be up to 30 characters.

```
IDENTIFICATION DIVISION.
PROGRAM-ID. HELLO.
PROCEDURE DIVISION.
DISPLAY 'Hello world '.
```

II. ENVIRONMENT DIVISION: it is an optional division, which describes the computer and other devices used to compile and execute the program and data files used in the program. It describes the system environment in which the program will run and specifies the input/output source required to run the program. It has two sections.
   - *CONFIGURATION SECTION*
   - *INPUT / OUTPUT SECTION*

- CONFIGURATION SECTION: it specifies info about the hardware on which the program is to be compiled and executed. It contains two paragraphs.
   - Source – machine: it refers to the system on which the program is compiled.
   - Object – machine: it refers to the system on which the program is executed.

All entries are written within the **AREA A** section.

- INPUT / OUTPUT SECTION: it specifies the file used or to be used by the program. It consists of two paragraphs:
  - File control: it gives information on the external dataset used in the program.
  - I / o control: it provides information about files used in the program.

```
ENVIRONMENT DIVISION.
CONFIGURATION SECTION.
 SOURCE-COMPUTER. XXX-IBM.
 OBJECT-COMPUTER. XXX-ZOS.

INPUT-OUTPUT SECTION.
 FILE CONTROL.
 SELECT FILEN ASSIGN TO DDNAME
 ORGANIZATION IS SEQUENTIAL.
```

III. DATA DIVISION: it is another optional division that takes of the description and definition of data item refers by the program. It includes their names, lengths, decimal point location, and storage formats. It takes care of the memory allocation required by the program.

<u>**DATA DIVISION CONSIST OF FOUR SECTIONS**</u>

- File section
- Working- storage section
- Linkage section
- The local – storage section

- File section: this subdivision shows the details of the records used in the file.it specifies the whole records structure of a file.
- Working – storage section: this subdivision specifies all temporary variables and the file structure that is used in the program.
- Linkage section: this subdivision specifies all data items received from external programs.
- Local – storage section: this subdivision takes care of all variables which are allotted and initialized during the program execution.

IV. PROCEDURE DIVISION: this is a mandatory division in a COBOL program. It contains the logic and executable statements of the program. Within this division, variables defined in the data division are implemented or initialized. This division must contain at least one statement. With the end statement being "stop run" which stops the program operation.

*   *Code sample*

```
IDENTIFICATION DIVISION.
   PROGRAM-ID. HELLO.


   DATA DIVISION.
      WORKING-STORAGE SECTION.
      01 WS-NAME PIC A (20).
      01 WS-ID PIC 9(5) VALUE 12345.


   PROCEDURE DIVISION.
      A000-FIRST-PARA.
         DISPLAY ".
         MOVE 'Obi Somu' TO WS-NAME.
         DISPLAY "My name is: "WS-NAME.
         DISPLAY "My ID is: "WS-ID.
      STOP RUN.
```

■ <u>SECTION</u>

It refers to the logical subdivisions of a COBOL program. And it is made up of paragraphs.

▪ PARAGRAPHS

It is a subdivision of either division or section.it is a user-defined variable followed by a period and it is usually consisting of zero or more statements.

POSITION   FIELD   DESCRIPTION

▪ SENTENCE

It is a combination of one or more statements. It must be ended with a period. The sentence appears only in the procedure division.

▪ STATEMENTS

It refers to the user-defined variabl.es., the conditional, loops, etc. written by the user which performs the task required of the program.

▪ CHARACTERS

It refers to the symbols and signs usable within the COBOL program. It is the lowest in the hierarchy and cannot be further divided.

COBOL CODING SHEET

1 to 6	Column Numbers	1 to 3 is reserved for a page number. 4 to 6 is reserved for line numbers.
7	Indicator	It can have (*, -, /). Asterisk (*) indicates comments, Hyphen (-) indicates continuation, Slash (/) indicate form feed.
8 to 11	Area A	All the COBOL divisions, sections, paragraphs, and special entries must start in Area A.
12 to 72	Area B	All COBOL statements should start in area B.
73 to 80	Identification Area	It can be used by the programmer as per the need.

For a COBOL program to execute it must be written in a compiler acceptable format. therefore, COBOL codes are written in a coding sheet that contains a total of 80-

COBOL CHARACTER SET

character position. The character position is grouped as follow;

A to Z	Alphabets in Upper case
a to z	Alphabets in Lower case
0 to 9	Numeric
+	Plus sign
*	Asterisk
	Space
-	Hyphen or minus
,	Comma
$	Currency sign
.	Period or Decimal point
;	Semicolon
(	Left parenthesis
)	Right parenthesis
"	Quotation marks
>	Greater than
<	Less than
'	Apostrophe
=	Equal sign
:	Colon

CHARACTER STRINGS

it refers to a collection of individual COBOL characters. It can be a:

- Comment
- Literal
- COBOL word

COMMENT

It is a COBOL character or character string which do not affect the program execution. It is of types:

- Comment line
- Comment entry

LITERAL

It is a constant that is directly hard-coded into a COBOL program. It is of two types:

- Alphanumeric literal
- Numeric literal
- Alphabetic literals

COBOL WORD

It is a character string that can be a reserved word or a user-defined word. Its length can be up to 30 characters.

Chapter 4

- COBOL DATA TYPE
- BASIC COBOL SYNTAX
- DATA LAYOUT

Data type being a crucial feature in other programming languages, so it is in COBOL programming. It is used to define the type of a variable or variable used in a program, thereby indicating the possible operation such a variable can perform.

In this language, the data division section of the program takes care of the definition of the variable required of the program.

TERMS FOR DATA DESCRIPTION IN COBOL

- Level number
- Data name
- Picture clause
- Value clause

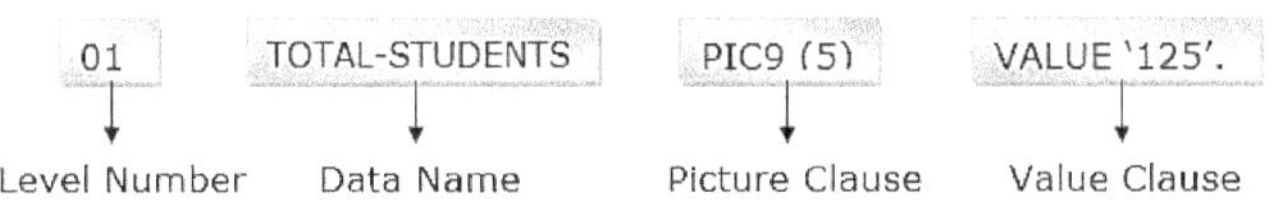

LEVEL NUMBER

It shows the level of data in a record. It differentiates between elementary and group items.

Sr.No.	Level Number & Description
1	**01** Record description entry
2	**02 to 49** Group and Elementary items
3	**66** Rename Clause items
4	**77** Items which cannot be sub-divided
5	**88** Condition name entry

DATA NAME

It is a mandatory entry, which must be done within the data division section before it can be used in the procedure division. Data names must have a user-defined variable.it gives reference to the memory location where actual data is stored. it can be an elementary or group

33

type depending on the arrangement. Reserved words must be used in the data name definition.

PICTURE CLAUSE

It is used to define numeric, alphanumeric, alphabet. It also takes care of signs; which can either + or -, decimal point positioning, length.

SYMBOLS OF PICTURE CLAUSE:

No.	Symbol & Description
1	**9** Numeric
2	**A** Alphabetic
3	**X** Alphanumeric

| 4 | **V** | Implicit Decimal |

| 5 | **S** | Sign |

| 6 | **P** | Assumed Decimal |

<u>VALUE CLAUSE</u>

It is a clause used in initializing data item, though, it is an optional clause it can encompass numerical literals, alphanumeric literals, and figurative constant which may be either a group or elementary items. entries initialized with the values clause should not exceed the length of the picture clause. Descriptive entries do not support redefines.

```cobol
IDENTIFICATION DIVISION.
PROGRAM-ID. HELLO.

DATA DIVISION.
 WORKING-STORAGE SECTION.
 01 WS-NUM1 PIC 99V9 VALUE IS 9.
 01 WS-NAME PIC A(6) VALUE 'Obi'.
 01 WS-ID PIC 99 VALUE 1.

PROCEDURE DIVISION.
 DISPLAY "WS-NUM1: "WS-NUM1.
 DISPLAY "WS-NAME: "WS-NAME.
 DISPLAY "WS-ID: "WS-ID.
STOP RUN.
```

Code sample on value clause

BASIC SYNTAX

As seen in all programming languages, the only way to achieve a working program is to follow the language syntax. COBOL Lang is made up of its syntax which includes;

- Character set
- Character string
- Character position in coding sheets

Well written COBOL program must follow the language programming standards for smooth execution. Newbie programmers are often encouraged to learn the basic syntax of a computer language for upon it is all other features are built upon.

BASIC VERB

COBOL verbs are used within the procedure division for data processing. It is the beginning word of a COBOL statement. There are numerous COBOL verbs and it includes the following:

- **INPUT / OUTPUT VERBS**
  - ACCEPT
  - DISPLAY
- **PROCESSING VERBS**
  - INITIALIZE
  - MOVE
  - ADD
  - SUBTRACT
  - MULTIPLY
  - DIVIDE
  - COMPUTE
- **STRING MANIPULATING VERBS**
  - STRING

- UNSTRING
  - INSPECT
- **FILE MANIPULATING VERBS**
  - OPEN
  - READ
  - WRITE
  - REWRITE
  - START
  - DELETE
- **PROGRAM SEGMENTING VERBS**
  - CALL
  - EXIT
  - EXIT PROGRAMS
  - GOT TO
  - PERFORM
  - STOP
  - STOP RUN.

DATA LAYOUT

It elaborates on the use of fields and their entry values. These description entries are as follows:

- Redefines entries
- Rename entries
- Usages clause

- Copy works

I. REDEFINES CLAUSE

It enables the use of the same storage location for multiply values. It defines given storage with different data description.

```
01 WS-OLD PIC X(10).
01 WS-NEW1 REDEFINES WS-OLD PIC 9(8).
01 WS-NEW2 REDEFINES WS-OLD PIC A(10).
```

Redefine syntax

NB:

- The level number of redefined and redefining items can not be the same and it can not be 66 or 88 level numbers.
- Using a values clause with a redefining item is wrong.

II. RENAMING CLAUSE

It is used to regroup data names and change their name. level number 66 is reserved for data renames. Do not rename data within the level number 01 or 77.

III. USAGE CLAUSE

This clause specifies the operating system in which the data is stored. This specification cannot be used with level number 66 or 88.

<u>USAGE CLAUSE OPTIONS;</u>

- Display
- Comp
- Com -1
- Comp - 2
- Comp - 3

IV. COPYWORKS:

It is like an inheritance function found in another computer language. It is a selection of code with a defined data structure that already existed in another part of the program, herby eliminating code repetition. It is achieved using the copy statement and must be entered within the working-storage section.

Chapter 5

- # LOOP STATEMENT
- # CONDITIONAL STATEMENTS

LOOP STATEMENTS

In computer programing, the need for a task continuous repetition often arises, a task like permutation, record reading, etc. in COBOL, to achieve the loop operation, the following are required:

- Perform thru
- Perform until
- Perform times
- Perform varying

I.   PERFORM THRU:

It executes a series of paragraphs by giving in the sequence first and the last name of the paragraph after the operation is done, the control returns. It is classified into:

- Inline perform
- Outline performs

```
PERFORM
  DISPLAY 'HELLO COBOL.!!!'
END-PERFORM.
```

II.   PERFORM UNTIL

In this operation, blocks of statements are executed continuously the until specified condition becomes true.

```
PERFORM A-PARA UNTIL COUNT=10
PERFORM A-PARA WITH TEST BEFORE
UNTIL COUNT=10
PERFORM A-PARA WITH TEST AFTER UNTIL
COUNT=10
```

III.   PERFORM TIMES

This loop statement performs repetitive execution of a paragraph according to a specified number of times.

```
PERFORM A-PARA 5 TIMES.
```

IV.   PERFORM VARYING

In this loop, a paragraph is executed until a given condition becomes true.

PERFORM A-PARA VARYING A FROM 1 BY 1 UNTIL A = 5.

CONDITIONAL STATEMENT:

In programming, conditional statements are used to change the program execution flow as coded by the coder. These statements include the following:

- If – else statement
- Relation statement
- Sign statement
- Class statement
- Negated statement

I. IF – ELSE STATEMENT

These conditional statements are executed based on a true or false condition. If a condition is true, the statement with the if block is executed, however, if the condition is false, then the code with the else block is executed.

The end – of the statement is required at the end i=of an if – statement block.

```
IF Condition
   {Statement Block}
[END-IF].
```

II.   RELATION STATEMENT

It is used in comparing two operands, which can be an identifier, literals, arithmetic expression.

III.   SIGN STATEMENT

It is used to check if a given numeric operand is greater than, equal to or zero.

```
[Data Name/Arithmetic Operation]

 [IS] [NOT]

[Positive, Negative or Zero]

[Data Name/Arithmetic Operation]
```

IV.   CLASS STATEMENT

It is checked whether an operand contains alphabet or numeric values.

<table>
<tr><td>

[Data Name/Arithmetic Operation>]

  [IS] [NOT]

[NUMERIC, ALPHABETIC, ALPHABETIC-LOWER, ALPHABETIC-UPPER]

[Data Name/Arithmetic Operation]

</td></tr>
</table>

V.   NEGATED STATEMENT:

It is used to convert a true value condition into a false value. it is achieved using the not keyword.

<table>
<tr><td>

IF NOT [CONDITION]
  Statements
END-IF.

</td></tr>
</table>

Chapter 6:

- STRING
- STRING FUNCTIONS.

STRING

It refers to non -numeric entries. Strings are an integral part of programming, for that there are functions in every programming language to handle and manipulate only strings.

STRING HANDLING FUNCTIONS:

In COBOL language, there are three functions for handling strings;

- Inspect
- String
- Unstring

I.   INSPECT

The inspect function is used to count/replace string character which can be alphabetic, alphanumeric, numeric characters. its operations involve two options.

* TALLYING – this option counts string characters
* REPLACE – this option replaces string characters.

II. STRING

These functions are used to concatenated two or more string. The delimited by clause is mandatory.

```
STRING ws-str1 DELIMITED BY SPACE
 ws-str2 DELIMITED BY SIZE
 INTO ws-destination-string
 WITH POINTER ws-count
 ON OVERFLOW DISPLAY message1
 NOT ON OVERFLOW DISPLAY message2
END-STRING
```

III. UNSTRING

It is another string handling function that is used to split one string into multiple substrings.it strongly requires the delimited by clause.

```
UNSTRING ws-str DELIMITED BY SPACE
INTO ws-str1, ws-str2
WITH POINTER ws-count
ON OVERFLOW DISPLAY message
NOT ON OVERFLOW DISPLAY message
END-UNSTRING.
```

Chapter 7

- # FILE HANDLING
- # FILE ORGANIZATION
- # FILE HANDLING VERB
- # FILE ACCESS MODES

FILE HANDLING

Just like most computer languages, COBOL supports file handling. however, a simple text file format is not supported. Rather,

- Physical sequential
- VSAM

The above is the two-file format supported by COBOL.

To understand more on the COBOL file handling system, some basic terms must be understood. It includes:

- Filed
- Record
- Physical record
- Logical records
- File records

I.   FIELD:

It indicates information stored about an element or object. These elements may be for instances, buyer id, buyer name, etc. The field is made up of attributes like:

- PRIMARY KEY:

It refers to a record identifying field that is unique to each record of the file.

- SECONDARY KEY:

It is another record identifying field that may be unique or not that is used to identify similar data.

- DESCRIPTION:

It is the field used in detailing a recorded element or entry.

II.   RECORD:

It is the collection of two or more files that details about a given entry.

III.   PHYSICAL RECORD:

It is a block of information that exists on an external device.

IV.   LOGICAL RECORD:

It refers to information being used by the program at the point of execution.

V.   FILE:

It refers to the organization of similar records.

FILE ORGANIZATION

It refers to ways by which files are organized to ensure smooth accessibly and efficiency during the file processing by the program. There are various way of organizing files and it includes;

- Sequential file organization
- Indexed sequential file organist

- Relative file organization

I. SEQUENTIAL FILE ORGANIZATION

In this file organization method, file ae stored and accessed in sequential order.

II. INDEXED SEQUENTIAL FILE ORGANIZATION;

In this method, the file is accessed sequentially but can also be accessed directly. this method is made up of two parts:

- data file
- index file

III. RELATIVE FILE ORGANIZATION:

In this file organization method, the file is organized according to their relative address.

FILE HANDLING VERBS

Performing operations on file in COBOL requires the use of COBOL file handling verbs which include, the following:

- Open
- Read
- Rewrite
- Rewrite
- Delete
- Start
- Close

COBOL ACCESS MODES

- **FILE ORGANIZATION:**

In file organization, as discussed previously each organization method uses a different access mode, which may be any of the below

- Random access
- Sequential access
- Dynamic access

The mode of access used by a given file organization determines how the file is being accessed by the program.

▪ **OTHER MODES**

➢ <u>OPEN VERB</u>:

It is one of the files handling verbs, which happens to be the first action to be performed on a file before a further operation can be executed on the file. it is the verb that creates awareness of the file to the program. However, it is made up of different modes that determine which operation can be performed on the file. these modes include;

- <u>INPUT</u>:

In this mode, the file can only be read and no other operation can be performed on it.

- <u>OUTPUT</u>:

In this mode, records can be inserted into the files. by overwriting the already existing records in the file.

- <u>EXTENDS</u>:

In this mode, records are inserted at the end of the file.

- <u>I/O</u>:

This mode allows the reading and rewriting of records in a file.

Chapter 8:

SUBROUTINES

In COBOL, subroutines are referred to as performing which can be written and compiled independently but cannot be executed independently. It is of two types;

- Internal subroutines
- External subroutines

The subroutines/subprograms play a very important role when the programmer was to program the same task in multiple sections for one program.

1. EXTERNAL SUBROUTINES:

It makes use of the calling verb to accomplish its operation which is transferring operation from the running program to the sub-program.

The program from which the calling verb makes the calls known as the main program while the program being called is known as the sub-program. After the subprogram finishes, the operation for which it was called the exit program is then used to transfer back the operation to the main program.

There Two Type of Calls in COBOL:

- Static call
- Dynamic calls

I.   STATICS CALLS: a sub-program called using this call method is to be loaded into the storage at compile time. this now means that the program is compiled with the NODYNAM compiler option.

II.   DYNAMIC CALL: when a sub-program is called using this option, the program is loaded into the storage during runtime. this is possible when the program has been compiled using the DYNAM and NODLL option.

For the main program to be able to call a sub-program the sub-program must contain the following required details:

- Linkage section
- Procedure division
- Exit program statement

Chapter 9:

- **AREAS NOT COVER**
- **COBOL KEYWORDS**
- **BONUS PROJECT**

AREAS NOT COVERED IN THIS MATERIAL

- Table processing.
- Internal sort.
- COBOL database handling.
- Advanced data processing.
- COBOL object-oriented programming (OOP) implementation.
- Array.
- COBOL Sequence Programming construction.
- Advanced Concept COBOL computes verb.
- Go to Statement.
- Array Subscribing.
- Array Indexing.

COBOL KEYWORDS

- ACCEPT
- ACCESS
- ADD
- ADDRESS [FOREIGN]
- ADVANCING
- AFTER

- ALL
- ALLOWING
- ALPHABET
- ALPHABETIC
- ALPHABETIC--LOWER
- ALPHABETIC--UPPER
- ALPHANUMERIC
- ALPHANUMERIC--EDITED
- ALSO
- ALTER
- ALTERNATE
- AND
- ANY
- APPLY
- ARE
- AREA
- AREAS
- ASCENDING
- ASSIGN
- AT
- AUTHOR
- AUTO [XOPEN]
- AUTOMATIC
- AUTOTERMINATE
- BACKGROUND-COLOR [XOPEN]
- BATCH
- BEFORE
- BEGINNING
- BELL [XOPEN]
- BINARY
- BINARY-CHAR [200X]
- BINARY-DOUBLE [200X]
- BINARY-LONG [200X]
- BINARY-SHORT [200X]
- BIT
- BITS
- BLANK
- BLINK [XOPEN]
- BLINKING
- BLOCK
- BOLD
- BOOLEAN
- BOTTOM
- BY
- CALL
- CANCEL
- CD

- CF
- CH
- CHANGED [FOREIGN]
- CHARACTER
- CHARACTERS
- CLASS
- CLOCK-UNITS
- CLOSE
- COBOL
- CODE
- CODE-SET
- COL [200X]
- COLLATING
- COLUMN
- COMMA
- COMMIT
- COMMON
- COMMUNICATION
- COMP
- COMP-1
- COMP-2
- COMP-3
- COMP-4
- COMP-5
- COMP-6
- COMP-X
- COMPUTATIONAL
- COMPUTATIONAL-1
- COMPUTATIONAL-2
- COMPUTATIONAL-3
- COMPUTATIONAL-4
- COMPUTATIONAL-5
- COMPUTATIONAL-6
- COMPUTATIONAL-X
- COMPUTE
- CONCURRENT
- CONFIGURATION
- CONNECT
- CONTAIN
- CONTAINS
- CONTENT
- CONTINUE
- CONTROL
- CONTROLS
- CONVERSION
- CONVERTING

- COPY
- CORE-INDEX [FOREIGN]
- CORR
- CORRESPONDING
- COUNT
- CRT
- CURRENCY
- CURRENT
- CURSOR
- DATA
- DATE
- DATE-COMPILED
- DATE-WRITTEN
- DAY
- DAY-OF-WEEK
- DB
- DB-ACCESS-CONTROL-KEY
- DB-CONDITION
- DB-CURRENT-RECORD-ID
- DB-CURRENT-RECORD-NAME
- DB-EXCEPTION
- Dbkey
- DB-RECORD-NAME
- DB-SET-NAME
- DB-STATUS
- DB-UWA
- DBCS [FOREIGN]
- DBKEY
- DE
- DEBUG-CONTENTS
- DEBUG-ITEM
- DEBUG-LENGTH
- DEBUG-LINE
- DEBUG-NAME
- DEBUG-NUMERIC-CONTENTS
- DEBUG-SIZE
- DEBUG-START
- DEBUG-SUB
- DEBUG-SUB-1
- DEBUG-SUB-2
- DEBUG-SUB-3
- DEBUG-SUB-ITEM
- DEBUG-SUB-N
- DEBUG-SUB-NUM

- DEBUGGING
- DECIMAL POINT
- DECLARATIVES
- DEFAULT
- DELETE
- DELIMITED
- DELIMITER
- DEPENDENCY
- DEPENDING
- DESCENDING
- DESCRIPTOR
- DESTINATION
- DETAIL
- DICTIONARY
- DISABLE
- DISCONNECT
- DISP [FOREIGN]
- DISPLAY
- DISPLAY-1 [FOREIGN]
- DISPLAY-6
- DISPLAY-7
- DISPLAY-9
- DIVIDE
- DIVISION
- DOES
- DOWN
- DUPLICATE
- DUPLICATES
- ECHO
- EDITING
- EGI
- EJECT [FOREIGN]
- ELSE
- EMI
- EMPTY
- ENABLE
- END
- END-ACCEPT
- END-ADD
- END-CALL
- END-COMMIT
- END-COMPUTE
- END-CONNECT
- END-DELETE
- END-DISCONNECT
- END-DIVIDE
- END-ERASE
- END-EVALUATE
- END-FETCH
- END-FIND
- END-FINISH
- END-FREE
- END-GET
- END-IF
- END-KEEP

- END-MODIFY
- END-MULTIPLY
- END-OF-PAGE
- END-PERFORM
- END-READ
- END-READY
- END-RECEIVE
- END-RECONNECT
- END-RETURN
- END-REWRITE
- END-ROLLBACK
- END-SEARCH
- END-START
- END-STORE
- END-STRING
- END-SUBTRACT
- END-UNSTRING
- END-WRITE
- ENDING
- ENTER
- ENTRY [FOREIGN]
- ENVIRONMENT
- EOL [XOPEN]
- EOP
- EOS [XOPEN]
- EQUAL
- EQUALS
- ERASE [XOPEN]
- ERROR
- ESI
- EVALUATE
- EVERY
- EXAMINE [FOREIGN]
- EXCEEDS
- EXCEPTION
- EXCLUSIVE
- EXHIBIT [FOREIGN]
- EXIT
- EXOR
- EXTEND
- EXTERNAL
- FAILURE
- FALSE
- FD
- FETCH
- FILE
- FILE CONTROL
- FILLER
- FINAL
- FIND
- FINISH
- FIRST
- FLOAT-

- EXTENDED [200X]
- FLOAT-LONG [200X]
- FLOAT-SHORT [200X]
- FOOTING
- FOR
- FOREGROUND-COLOR [XOPEN]
- FREE
- FROM
- FULL [XOPEN]
- FUNCTION
- GENERATE
- GET
- GIVING
- GLOBAL
- GO
- GOBACK [FOREIGN]
- GREATER
- GROUP
- HEADING
- HIGH-VALUE
- HIGH-VALUES
- HIGHLIGHT [XOPEN]
- I-O
- I-O-CONTROL
- ID [FOREIGN]
- IDENT
- IDENTIFICATION
- IF
- IN
- INCLUDING
- INDEX
- INDEXED
- INDICATE
- INITIAL
- INITIALIZE
- INITIATE
- INPUT
- INPUT-OUTPUT
- INSPECT
- INSTALLATION
- INTO
- INVALID
- IS
- JUST
- JUSTIFIED
- KANJI [FOREIGN]
- KEEP
- KEY
- LABEL
- LAST

- LD
- LEADING
- LEFT
- LENGTH
- LESS
- LIMIT
- LIMITS
- LINAGE
- LINAGE-COUNTER
- LINE
- LINE-COUNTER
- LINES
- LINKAGE
- LOCALLY
- LOCK
- LOCK-HOLDING
- LOW-VALUE
- LOW-VALUES
- LOWLIGHT [XOPEN]
- MANUAL
- MATCH
- MATCHES
- MEMBER
- MEMBERSHIP
- MEMORY
- MERGE
- MESSAGE
- MODE
- MODIFY
- MODULES
- MOVE
- MULTIPLE
- MULTIPLY
- NAMED [FOREIGN]
- NATIVE
- NEGATIVE
- NEXT
- NO
- NON-NULL
- NOT
- NOTE [FOREIGN]
- NULL
- NUMBER
- NUMERIC
- NUMERIC-EDITED
- OBJECT-COMPUTER
- OCCURS
- OF
- OFF
- OFFSET
- OMITTED
- ON

- ONLY
- OPEN
- OPTIONAL
- OPTIONS [200X]
- OR
- ORDER
- OTHERWISE [FOREIGN]
- PACKED-DECIMAL
- PADDING
- PAGE
- PAGE-COUNTER
- PASSWORD [FOREIGN]
- PERFORM
- PF
- PH
- PIC
- PICTURE
- PLUS
- POINTER
- POSITION
- POSITIONING [FOREIGN]
- POSITIVE
- PREVIOUS
- PRINTING
- PRIOR
- PROCEDURE
- PROCEDURES
- PROCEED
- PROGRAM
- PROGRAM-ID
- PROTECTED
- PURGE
- QUEUE
- QUOTE
- QUOTES
- RANDOM
- RD
- READ
- READERS
- READY
- REALM
- REALMS
- RECEIVE
- RECONNECT
- RECORD
- RECORD-NAME
- RECORD-OVERFLOW [FOREIGN]
- RECORDING [FOREIGN]
- RECORDS
- REDEFINES

- REEL
- REFERENCE
- REFERENCE-MODIFIER
- REFERENCES
- REGARDLESS
- RELATIVE
- RELEASE
- RELOAD [FOREIGN]
- REMAINDER
- REMARKS [FOREIGN]
- REMOVAL
- RENAMES
- REORG-CRITERIA [FOREIGN]
- REPLACE
- REPLACING
- REPORT
- REPORTING
- REPORTS
- REQUIRED [XOPEN]
- RERUN
- RESERVE
- RESET
- RETAINING
- RETRIEVAL
- RETURN
- RETURN-CODE [XOPEN]
- RETURNING [FOREIGN]
- REVERSE-VIDEO [XOPEN]
- REVERSED
- REWIND
- REWRITE
- RF
- RH
- RIGHT
- RMS-CURRENT-FILENAME
- RMS-CURRENT-STS
- RMS-CURRENT-STV
- RMS-FILENAME
- RMS-STS
- RMS-STV
- ROLLBACK
- ROUNDED
- RUN
- SAME
- SCREEN [XOPEN]

- SD
- SEARCH
- SECTION
- SECURE [XOPEN]
- SECURITY
- SEGMENT
- SEGMENT-LIMIT
- SELECT
- SEND
- SENTENCE
- SEPARATE
- SEQUENCE
- SEQUENCE-NUMBER
- SEQUENTIAL
- SERVICE [FOREIGN]
- SET
- SETS
- SIGN
- SIGNED [200X]
- SIZE
- SKIP1 [FOREIGN]
- SKIP2 [FOREIGN]
- SKIP3 [FOREIGN]
- SORT
- SORT-MERGE
- SOURCE
- SOURCE-COMPUTER
- SPACE
- SPACES
- SPECIAL-NAMES
- STANDARD
- STANDARD-1
- STANDARD-2
- START
- STATUS
- STOP
- STORE
- STREAM
- STRING
- SUB-QUEUE-1
- SUB-QUEUE-2
- SUB-QUEUE-3
- SUB-SCHEMA
- SUBTRACT
- SUCCESS
- SUM
- SUPPRESS
- SYMBOLIC

- SYNC
- SYNCHRONIZED
- TABLE
- TALLYING
- TAPE
- TERMINAL
- TERMINATE
- TEST
- TEXT
- THAN
- THEN
- THROUGH
- THRU
- TIME
- TIMES
- TO
- TOP
- TRACE [FOREIGN]
- TRAILING
- TRANSFORM [FOREIGN]
- TRUE
- TYPE
- UNDERLINE [XOPEN]
- UNDERLINED
- UNEQUAL
- UNIT
- UNLOCK
- UNSIGNED [200X]
- UNSTRING
- UNTIL
- UP
- UPDATE
- UPDATERS
- UPON
- USAGE
- USAGE-MODE
- USE
- USING
- VALUE
- VALUES
- VARYING
- VFU-CHANNEL
- WAIT
- WHEN
- WHERE
- WITH
- WITHIN
- WORDS
- WORKING-STORAGE
- WRITE
- WRITERS
- ZERO
- ZEROS

* ZEROS

BONUS PROJECT

```
**************************************************
*****************
* Author: Obi Somu
* Date: September 21, 2020
* Purpose: Book Example Application
**************************************************
*****************
IDENTIFICATION DIVISION.

PROGRAM-ID. CALCULATOR.

DATA DIVISION.

WORKING-STORAGE SECTION.

77 NUM_1 PIC 9(4).

77 NUM_2 PIC 9(4).

77 SOLVE_SUM PIC 9(4).

77 SOLVE_DIFF PIC 9(4).

77 SOLVE_PRODUCT PIC 9(4).

77 SOLVE_QUOTIENT PIC 9(4).

PROCEDURE DIVISION.
```

```
PARA.

DISPLAY "MATH CALCULATOR IN COBOL".

DISPLAY " ".

DISPLAY "ENTER THE FIRST VALUE: ".

ACCEPT NUM_1.

DISPLAY "ENTER THE SECOND VALUE: ".

ACCEPT NUM_2.

COMPUTE SOLVE_SUM = NUM_1 + NUM_2.

COMPUTE SOLVE_DIFF = NUM_1 - NUM_2.

COMPUTE SOLVE_PRODUCT = NUM_1 * NUM_2.

COMPUTE SOLVE_QUOTIENT = NUM_1 / NUM_2.

DISPLAY " ".

DISPLAY "===== DISPLAY RESULT =====".

DISPLAY " ".

DISPLAY "THE SUM IS ".

DISPLAY SOLVE_SUM.

DISPLAY " ".

DISPLAY "THE DIFFERENCE IS ".

DISPLAY SOLVE_DIFF.
```

```
DISPLAY " ".

DISPLAY "THE PRODUCT IS ".

DISPLAY SOLVE_PRODUCT.

DISPLAY " ".

DISPLAY "THE QUOTIENT IS ".

DISPLAY SOLVE_QUOTIENT.

STOP RUN.
```

Chapter 10

- COBOL INTERVIEW QUESTIONS
- COMPUTER GLOSSARY

COBOL INTERVIEW QUESTION

1. Tell me what you know about COBOL.
2. Describes the most important features of COBOL.
3. Explain the difference between subscript and index.
4. How do COBOL's search functions differ?
5. Why would you use a scope terminator?
6. Explain the file opening modes used in COBOL.
7. Why would you want to perform a paragraph instead of a section?
8. Explain the process of writing into a file.
9. Explain the difference between a static and a dynamic call.
10. How do you perform a binary search?
11. Explain the difference between COMP-1 and COMP-2.
12. When would you use a compute statement?
13. Why should you know the difference between using TEST BEFORE and TEST AFTER?
14. What are the consequences of coding GO BACK instead of STOP RUN?

15. In which division is the FILE CONTROL paragraph located?
16. Which verb tells the program to update a file?
17. When you have two tables, which one will the SEARCH function use?
18. What are some examples of command terminators?
19. What does length mean to COBOL?
20. When operating a sequential file, which mode would you use?
21. What does the REPLACING option do in a copy statement?
22. What is the purpose of the LINKAGE SECTION?
23. How are arrays defined in COBOL?

COMPUTER GLOSSARY

1. **APPLICATION**: Computer software that performs a task or set of tasks.
2. **BINARY CODE**: The most basic language a computer understands.
3. **BIT**: The smallest piece of computer information, either the number 0 or 1.
4. **BOOT**: To start up a computer.

5. **BROWSER**: Software used to navigate the Internet.
6. **BUG**: A malfunction due to an error in the program.
7. **BYTE**: Most computers use combinations of eight bits, called bytes, to represent one character of data or instructions.
8. **CACHE:** A small data-memory storage area that a computer can use to instantly re-access data instead of re-reading the data from the source, such as a hard drive.
9. **CD-ROM**: *Compact Disc Read-Only Memory,* an optically read disc designed to hold information such as music, reference materials, or computer software. A single CD-ROM can hold around 640 megabytes of data, enough for several encyclopedias. Most software programs are now delivered on CD-ROMs.
10. **CGI: Common Gateway Interface,** a programming standard that allows visitors to fill out form fields on a Web page and have that information interact with a database, possibly coming back to the user as another Web page.
11. **Chat** Typing text into a message box on a screen to engage in dialogue with one or more people via the Internet or other network.

12. **CHIP:** A tiny wafer of silicon-containing miniature electric circuits that can store millions of bits of information.
13. **CLIENT:** A single user of a network application that is operated from a server.
14. **COOKIE**: A text file sent by a Web server that is stored on the hard drive of a computer and relays back to the Web server things about the user, his or her computer, and/or his or her computer activities.
15. **CPU:** Central Processing Unit. The brain of the computer.
16. **CRACKER**: A person who breaks into a computer through a network, without authorization, and with mischievous or destructive intent.
17. **CRASH:** A hardware or software problem that causes information to be lost or the computer to malfunction. Sometimes a crash can cause permanent damage to a computer.
18. **CURSOR:** A moving position-indicator displayed on a computer monitor that shows a computer operator where the next action or operation will take place.
19. **CYBERSPACE:** Slang for internet i.e. An international conglomeration of interconnected computer networks.
20. **DATABASE:** A collection of similar information stored in a file, such as a

database of addresses. This information may be created and stored in a database management system (DBMS).

21. **DEBUG:** Slang. To find and correct equipment defects or program malfunctions.

22. **DEFAULT:** The pre-defined configuration of a system or an application. In most programs, the defaults can be changed to reflect personal preferences.

23. **DESKTOP:** The main directory of the user interface. Desktops usually contain icons that represent links to the hard drive, a network (if there is one), and trash or recycling can for files to be deleted.

24. **DESKTOP PUBLISHING:** The production of publication-quality documents using a personal computer in combination with text, graphics, and page layout programs.

25. **DIRECTORY:** A repository where all files are kept on the computer.

26. **DISK DRIVE:** The equipment that operates a hard or floppy disc.

27. **DOMAIN:** Represents an IP (Internet Protocol) address or set of IP addresses that comprise a domain. Each domain name ends with a suffix that indicates what top-level domain it belongs to. These are:.com for commercial,.gov for

government,.org for the organization, Edu for educational institution,.biz for business,.info for information,.tv for television,.ws for a website. Domain suffixes may also indicate the country in which the domain is registered. No two parties can ever hold the same domain name.

28. **DOMAIN NAME:** The name of a network or computer linked to the Internet. Domains are defined by a common IP address or set of similar IP addresses.
29. **DOWNLOAD:** The process of transferring information from a web site (or other remote location on a network) to the computer. It is possible to download a file that includes text, image, audio, video, and many others.
30. **DOS:** Disk Operating System. An operating system designed for early IBM-compatible PCs.
31. **DROP-DOWN MENU**: a window that opens vertically on-screen to display context-related options. Also called the pop-up menu or the pull-down menu.
32. **DSL:** *Digital Subscriber Line*, a method of connecting to the Internet via a phone line. A DSL connection uses copper telephone lines but can relay data at much higher speeds than modems and does not interfere with telephone use.

33. **DVD:** Digital Video Disc. Like a CD-ROM, it stores and plays both audio and video.
34. **E-BOOK:** An electronic (usually hand-held) reading device that allows a person to view digitally stored reading materials.
35. **EMAIL:** Electronic mail; messages, including memos or letters, are sent electronically between networked computers that may be across the office or around the world.
36. **EMOTICON:** A text-based expression of emotion created from ASCII characters that mimic a facial expression.
37. **ENCRYPTION:** The process of transmitting scrambled data so that only authorized recipients can unscramble it.
38. **ETHERNET:** A type of network.
39. **ETHERNET CARD:** A board inside a computer to which a network cable can be attached.
40. **FILE:** A set of data that is stored in the computer.
41. **FIREWALL:** A set of security programs that protect a computer from outside interference or access via the Internet.
42. **FOLDER:** A structure for containing electronic files. In some operating systems, it is called a directory.

43. **FONTS:** Sets of typefaces (or characters) that come in different styles and sizes.
44. **FREEWARE:** Software created by people who are willing to give it away for the satisfaction of sharing or knowing they helped to simplify other people's lives. It may be free-standing software, or it may add functionality to existing software.
45. **FTP:** *File Transfer Protocol,* a format and set of rules for transferring files from a host to a remote computer.
46. **GIGABYTE (GB):**1024 megabytes. Also called a gig.
47. **GLITCH:** The cause of an unexpected malfunction.
48. **GOPHER:** An Internet search tool that allows users to access textual information through a series of menus, or if using FTP, through downloads.
49. **GUI:** *Graphical User Interface,* a system that simplifies selecting computer commands by enabling the user to point to symbols or illustrations (called icons) on the computer screen with a mouse.
50. **GROUPWARE:** Software that allows networked individuals to form groups and collaborate on documents, programs, or databases.
51. **HACKER:** A person with technical expertise who experiments with

computer systems to determine how to develop additional features. Hackers are occasionally requested by system administrators to try and break into systems via a network to test security. The term hacker is sometimes incorrectly used interchangeably with crackers. A hacker is called a white hat and a cracker a black hat.

52. **HARD COPY:** A paper printout of what you have prepared on the computer.

53. **HARD DRIVE:** Another name for the hard disc that stores information on a computer.

54. **HARDWARE:** The physical and mechanical components of a computer system, such as the electronic circuitry, chips, monitor, disks, disk drives, keyboard, modem, and printer.

55. **HOME PAGE:** The main page of a Web site used to greet visitors, provide information about the site, or to direct the viewer to other pages on the site.

56. **HTML:** *Hypertext Markup Language,* a standard of text markup conventions used for documents on the World Wide Web. Browsers interpret the codes to give the text structure and formatting (such as bold, blue, or italic).

57. **HTTP:** *Hypertext Transfer Protocol,* a common system used to request and send HTML documents on the World

Wide Web. It is the first portion of all URL addresses on the World Wide Web.

58. **HTTPS:** *Hypertext Transfer Protocol Secure,* often used in intracompany internet sites. Passwords are required to gain access.

59. **HYPERLINK:** Text or an image that is connected by hypertext coding to a different location. By selecting the text or image with a mouse, the computer jumps to (or displays) the linked text.

60. **HYPERMEDIA:** Integrates audio, graphics, and/or video through links embedded in the main program.

61. **HYPERTEXT:** A system for organizing text through links, as opposed to a menu-driven hierarchy such as Gopher. Most Web pages include hypertext links to other pages at that site, or to other sites on the World Wide Web.

62. **ICONS:** Symbols or illustrations appearing on the computer screen that indicate program files or other computer functions.

63. **INPUT:** Data that goes into a computer device.

64. **INPUT DEVICE:** A device, such as a keyboard, stylus and tablet, mouse, puck, or microphone, that allows input of information (letters, numbers, sound, video) to a computer.

65. **INSTANT MESSAGING (IM):** A chat application that allows two or more people to communicate over the Internet via real-time keyed-in messages.
66. **INTERFACE:** The interconnections that allow a device, a program, or a person to interact.
67. **INTERNET:** An international conglomeration of interconnected computer networks.
68. **IP ADDRESS:** An Internet Protocol address is a unique set of numbers used to locate another computer on a network.
69. **JAVA:** An object-oriented programming language designed specifically for programs to be used over the Internet. Java allows programmers to create small programs or applications (applets) to enhance Web sites.
70. **JAVASCRIPT/ECMA SCRIPT:** A programming language used almost exclusively to manipulate content on a web page.
71. **KILOBYTE (KB):** Equal to 1,024 bytes.
72. **LINUX:** A UNIX - like, an open-source operating system developed primarily by Linus Torvalds. Linux is free and runs on many platforms, including both PCs and Macintoshes.

73. **LAPTOP AND NOTEBOOK**: Small, lightweight, portable battery-powered computers that can fit onto your lap.
74. **MACRO:** A script that operates a series of commands to perform a function. It is set up to automate repetitive tasks.
75. **MAC OS:** An operating system with a graphical user interface, developed by Apple for Macintosh computers.
76. **MEGABYTE (MB):** Equal to 1,048,576 bytes, usually rounded off to one million bytes (also called a meg).
77. **MEMORY:** Temporary storage for information, including applications and documents. The information must be stored to a permanent device, such as a hard disc or CD-ROM before the power is turned off, or the information will be lost.
78. **MENU:** A context-related list of options that users can choose from.
79. **MENU BAR:** The horizontal strip across the top of an application's window. Each word on the strip has a context-sensitive drop-down menu containing features and actions that are available for the application in use.
80. **MERGE:** To combine two or more files into a single file.
81. **MHZ:** An abbreviation or *Megahertz,* or *one million hertz.* One MHz represents one million clock cycles per second and is the

measure of a computer microprocessor's speed.

82. **MICROPROCESSOR:** A complete central processing unit (CPU) contained on a single silicon chip.

83. **MONITOR:** A video display terminal.

84. **MOUSE:** A small hand-held device, like a trackball, used to control the position of the cursor on the video display; movements of the mouse on a desktop correspond to movements of the cursor on the screen.

85. **MP3:** Compact audio and video file format. The small size of the files makes it easy to download and email. The format is used in portable playback devices.

86. **MULTIMEDIA:** Software programs that combine text and graphics with sound, video, and animation. A multimedia PC contains the hardware to support these capabilities.

87. **MS-DOS:** An early operating system developed by Microsoft Corporation (Microsoft Disc Operating System).

88. **NETWORK:** A system of interconnected computers.

89. **OPEN-SOURCE:** Computer programs whose source code was revealed to the general public so that it could be developed openly. Software licensed as open-source can be freely changed or adapted to new uses, meaning that the source code of the operating system is freely available to the public.

90. **PC:** A single user computer containing a CPU and one or more memory circuits.